Skeletal Muscle & Muscular Dystrophy

A Visual Approach

Colloquium Series on the Cell Biology of Medicine

Editor
Joel D. Pardee, Weill Cornell Medical College

Published

Skeletal Muscle & Muscular Dystrophy: A Visual Approach
Donald A. Fischman
2009

How the Heart Develops: A Visual Approach
Donald A. Fischman
2009

Forthcoming Books

The Body Plan: How Structure Creates Function
Joel D. Pardee
2009

Bones: Growth, Strength, and Osteoporosis
Michelle Fuortes
2009

Breast Cancer and the Estrogen Receptor
Joel D. Pardee
2009

(Forthcoming Books continued on page 49)

Copyright © 2009 by Morgan & Claypool Life Sciences

All rights reserved. No part of this publication may be reproduced, stored in a retrieval system, or transmitted in any form or by any means—electronic, mechanical, photocopy, recording, or any other except for brief quotations in printed reviews, without the prior permission of the publisher.

Skeletal Muscle & Muscular Dystrophy: A Visual Approach
Donald A. Fischman
www.morganclaypool.com

ISBN: 9781615040032 paperback

ISBN: 9781615040049 ebook

DOI: 10.4199/C00001ED1V01Y200904CBM002

A Publication in the Morgan & Claypool Life Sciences series

COLLOQUIUM SERIES ON THE CELL BIOLOGY OF MEDICINE

Book #2

Series Editor: Joel D. Pardee, Weill Cornell Medical College

Series ISSN TBD

Skeletal Muscle & Muscular Dystrophy

A Visual Approach

Donald A. Fischman
Weill Cornell Medical College

COLLOQUIUM SERIES ON THE CELL BIOLOGY OF MEDICINE #2

MORGAN & CLAYPOOL LIFE SCIENCES

ABSTRACT

Histologically, muscle is conveniently divided into two groups, striated and nonstriated, based on whether the cells exhibit cross-striations in the light microscope (Figure 3). Smooth muscle is involuntary—its contraction is controlled by the autonomic nervous system. Striated muscle includes both cardiac (involuntary) and skeletal (voluntary). The former is innervated by visceral efferent fibers of the autonomic nervous system, whereas the latter is innervated by somatic efferent fibers, most of which have their cell bodies in the ventral, motor horn of the spinal cord. Smooth muscle is designed to have slow, relatively sustained contractions, while striated muscle contracts rapidly and usually phasically.

Both cardiac and smooth muscle cells are mononucleated, whereas skeletal muscle cells (fibers) are multinucleated. [In aging hearts or hypertrophied hearts, cardiac muscle cells are often binucleated.] Multinucleation of skeletal muscle arises during development by the cytoplasmic fusion of muscle precursor cells, myoblasts. Adult skeletal muscle cells do not divide; that is also true of most cardiac myocytes. However, skeletal muscle exhibits a considerable amount of regeneration after injury. This is because adult skeletal muscle contains a stem cell, the satellite cell, which lies beneath the basement membrane surrounding the muscle fibers. [The multinucleation of cardiac muscle arises from karyokinesis without cytokinesis.]

A diagrammatic series of enlargements of skeletal muscle are shown in Figure 4. A bundle of muscle fibers (fasciculus) is cut from the deltoid muscle. Each muscle cell is termed a myofiber or muscle fiber. Each muscle fiber contains contractile organelles termed myofibrils, which contain the contractile units of muscle termed sarcomeres. The sarcomeres are composed of myofilaments, which in turn are composed of contractile proteins.

Muscle connective tissue layers are organized in concentric layers that are important in the entry and exit of vessels and nerves to and from the tissue. These are shown in Figure 5.

The outermost layer is the epimysium or muscle sheath. Connective tissue septae (perimysium) run radially into the muscle tissue, dividing it into muscle fascicles. The deepest layer, surrounding each of the muscle fibers is the endomysium. The endomysium is in direct contact with a basal lamina that ensheathes each muscle fiber. It surrounds the plasma membrane of the muscle fiber termed the sarcolemma.

KEYWORDS

muscle cell (myofiber), sarcomere, myosin molecule, regulation of muscle contraction, sarcoplasmic reticulum, muscle fiber types, myotendon junction, satellite cell, muscular dystrophy

Contents

Introduction to the Cell Biology of Medicine

In order to learn, we must be able to remember, and in the world of science and medicine, we remember what we envision, not what we hear. It is with this essential precept in mind that we offer the Cell Biology of Medicine lecture series. Each lecture is given by faculty accomplished in teaching the scientific basis of disease to both graduate and medical students. In this modern age, it has become clear that everyone is vastly interested in how our bodies work and what has gone wrong in a disease. It is likewise evident that the only way to understand medicine is to engrave in our mind's eye a clear vision of the biological processes that give us the gift of life. In these lectures, we are dedicated to holding up for the viewer an insight into the biology behind the body. Each lecture demonstrates cell, tissue, and organ function in health and disease, and it does so in a visually striking style. Left to its own devices, the mind will quite naturally remember the pictures. Enjoy the show.

Joel Pardee
New York, NY

Introduction

Historically, the modern study of muscle anatomy can be dated to the Belgian physician–anatomist Andreas Vesalius (1514–1664) who began modern gross anatomy with his dissections in Padua, Italy (Figure 1). His work was published in the classic *De Humani Corporis Fabrica*. The great drawings in that work were done in collaboration with Titian's atelier working in Venice. His portrait is shown with one of the drawings (Figure 2).

Histologically, muscle is conveniently divided into two groups, striated and nonstriated, based on whether the cells exhibit cross-striations in the light microscope (Figure 3). Smooth muscle is involuntary—its contraction is controlled by the autonomic nervous system. Striated muscle includes both cardiac (involuntary) and skeletal (voluntary). The former is innervated by visceral efferent fibers of the autonomic nervous system, whereas the latter is innervated by somatic efferent fibers, most of which have their cell bodies in the ventral, motor horn of the spinal cord. Smooth muscle is designed to have slow, relatively sustained contractions, while striated muscle contracts rapidly and usually phasically.

Both cardiac and smooth muscle cells are mononucleated, whereas skeletal muscle cells (fibers) are multinucleated. [In aging hearts or hypertrophied hearts, cardiac muscle cells are often binucleated.] Multinucleation of skeletal muscle arises during development by the cytoplasmic fusion of muscle precursor cells, myoblasts. Adult skeletal muscle cells do not divide; that is also true of most cardiac myocytes. However, skeletal muscle exhibits a considerable amount of regeneration after injury. This is because adult skeletal muscle contains a stem cell, the satellite cell, which lies beneath the basement membrane surrounding the muscle fibers. [The multinucleation of cardiac muscle arises from karyokinesis without cytokinesis.]

A diagrammatic series of enlargements of skeletal muscle are shown in Figure 4. A bundle of muscle fibers (fasciculus) is cut from the deltoid muscle. Each muscle cell is termed a myofiber or muscle fiber. Each muscle fiber contains contractile organelles termed myofibrils, which contain the contractile units of muscle termed sarcomeres. The sarcomeres are composed of myofilaments, which in turn are composed of contractile proteins.

Muscle connective tissue layers are organized in concentric layers that are important in the entry and exit of vessels and nerves to and from the tissue. These are shown in Figure 5.

The outermost layer is the epimysium or muscle sheath. Connective tissue septae (perimysium) run radially into the muscle tissue, dividing it into muscle fascicles. The deepest layer, surrounding each of the muscle fibers is the endomysium. The endomysium is in direct contact with a basal lamina that ensheathes each muscle fiber. It surrounds the plasma membrane of the muscle fiber termed the sarcolemma.

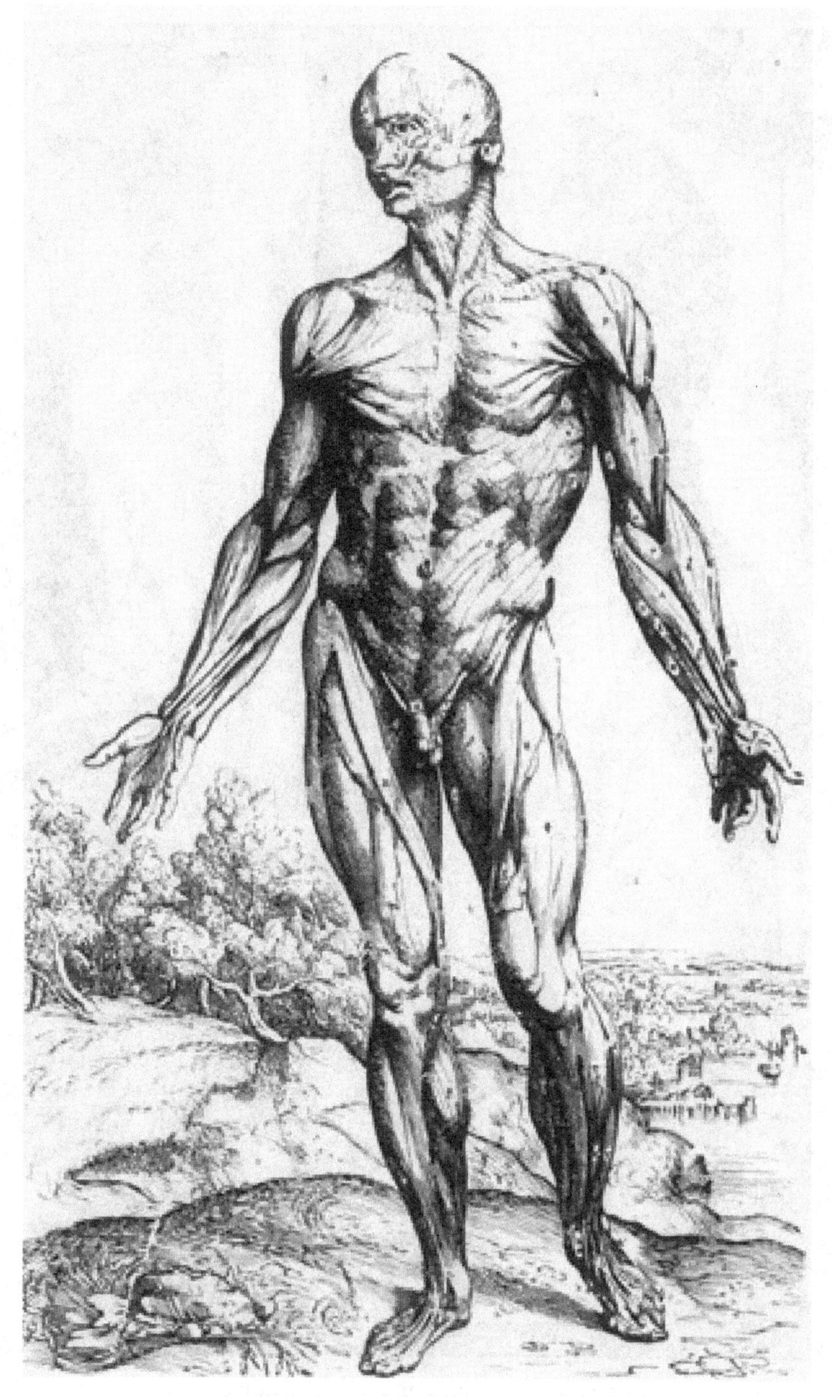

FIGURE 1: Andreas Vesalius illustration of muscle anatomy. Permission pending.

FIGURE 2: Woodcut portrait of Andreas Vesalius. Permission pending.

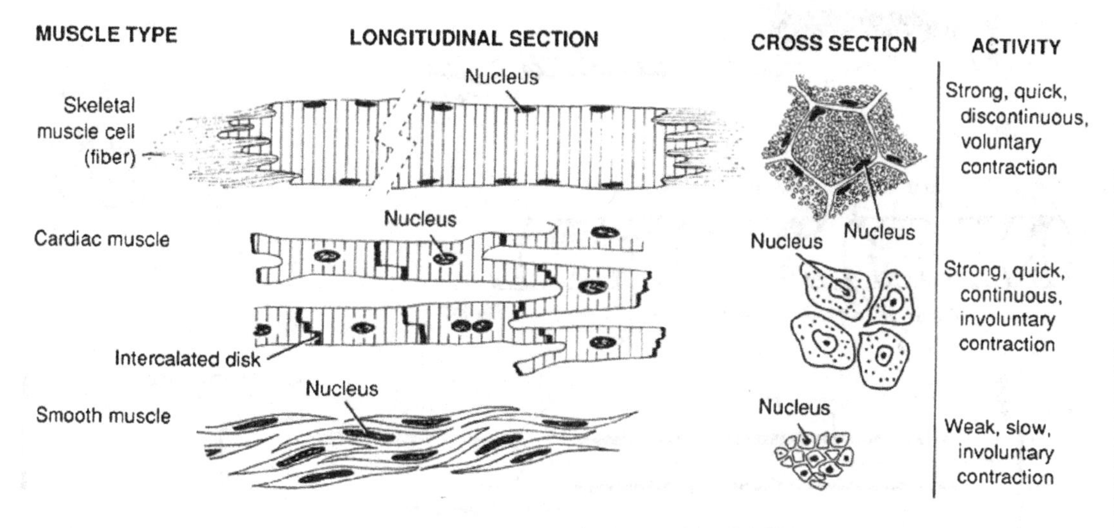

FIGURE 3: Types of muscle. Permission pending.

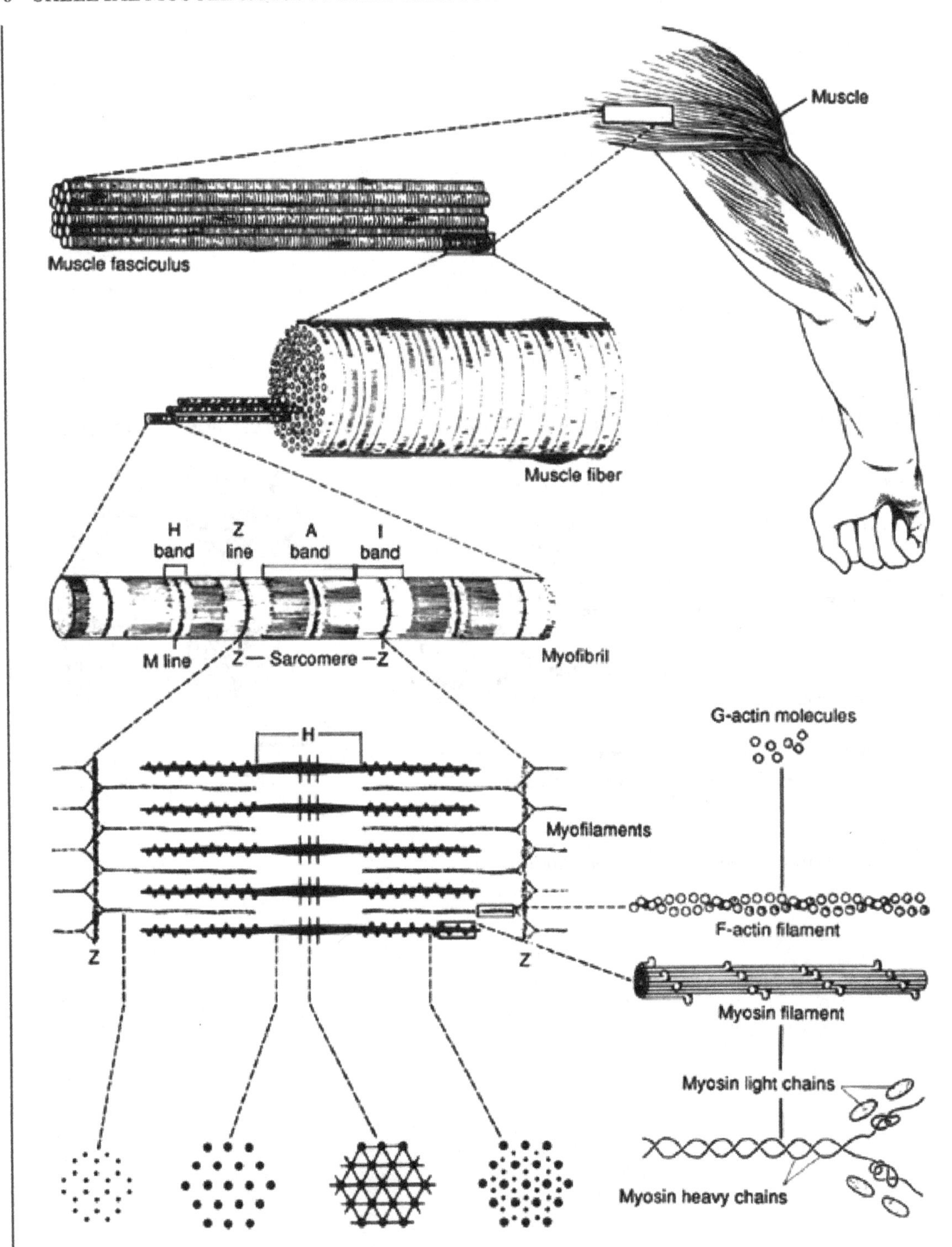

FIGURE 4: Diagrammatic series of enlargements of skeletal muscle. Permission pending.

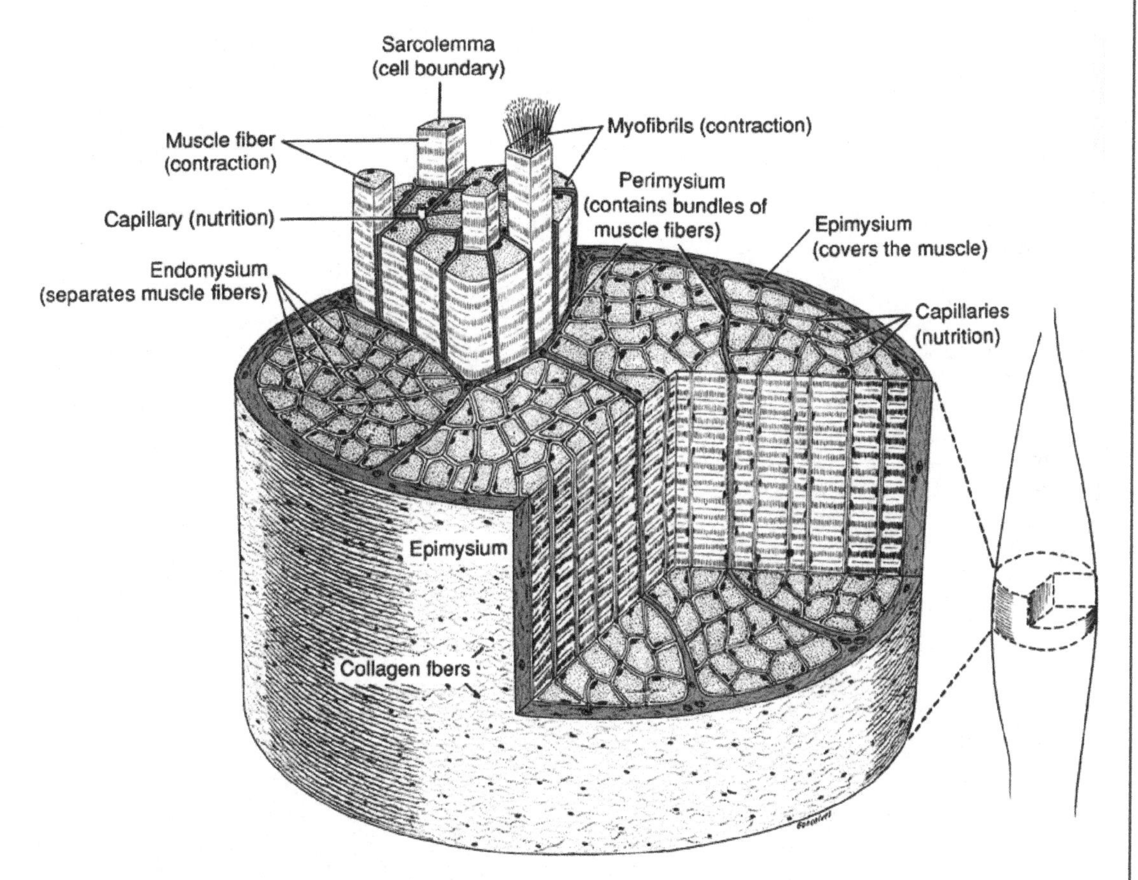

FIGURE 5: Muscle connective tissue layers. Permission pending.

The Muscle Cell (Myofiber)

Myofibers are cylindrical cells ~75–100 μm in diameter and of variable length, usually many millimeters long. The ends terminate in dense connective attachments, e.g., the myotendon junction (see below). A longitudinal section of skeletal muscle is presented in Figure 6. This paraffin section has been stained with hematoxylin–eosin. The transverse striations are evident. Muscle fibers are typically eosinophilic because of their low content of either free or membrane-bound ribosomes. The nuclei are peripherally distributed. The bulk of the sarcoplasm is filled with myofibrils. Interspersed between the myofibrils and beneath the sarcolemma are mitochondria. Their number varies depending on muscle fiber type. An extensive smooth endoplasmic reticulum, termed the sarcoplasmic reticulum (SR), surrounds the myobibrils.

A cross section of skeletal muscle is presented in Figure 7. In this figure, the peripheral disposition of the myonuclei is clearly evident. Centrally located nuclei in adult muscle are an indication of muscle injury followed by regeneration. Note that some of the fibers are more eosinophilic than others; this is an indication of the different fiber types in a muscle. This will be discussed below. Only about half of the nuclei seen in muscle are myonuclei (also true of the myocardium). Many are fibroblasts and endothelial cells and a small fraction are satellite cells (myogenic stem cells) to be discussed below.

Each muscle fiber receives motor nerve innervation at a specialized structure termed the neuromuscular junction (Figure 8). It is at this site that synaptic release of acetylcholine occurs, resulting in depolarization of the muscle plasma membrane and generation of the muscle action potential.

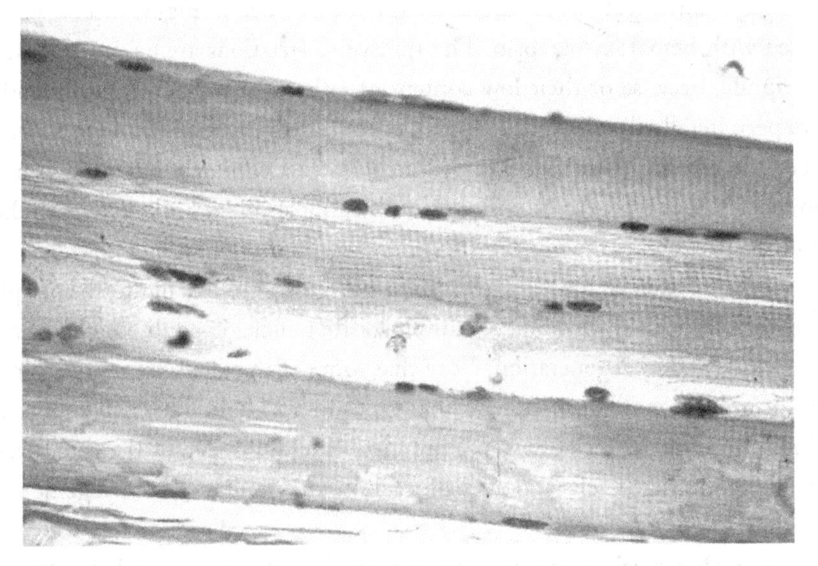

FIGURE 6: Light micrograph of mammalian skeletal muscle. Permission pending.

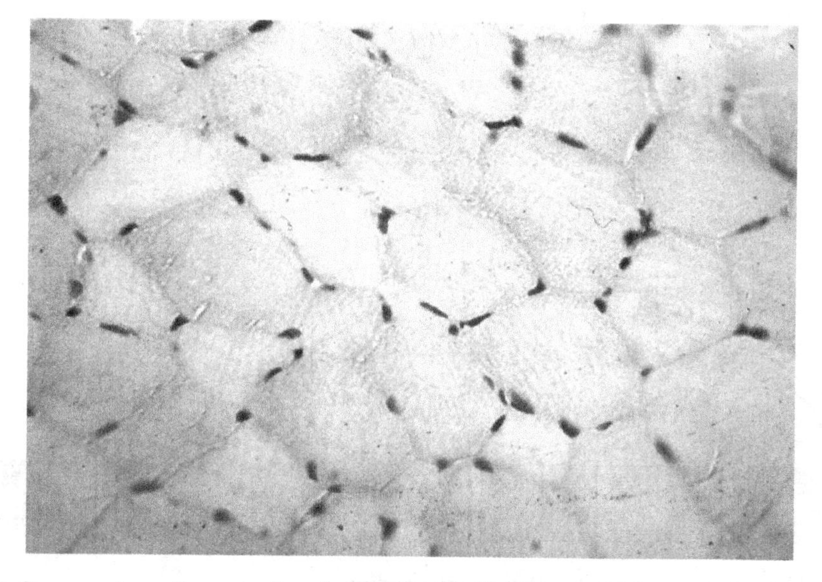

FIGURE 7: Cross section of mammalian skeletal muscle. Permission pending.

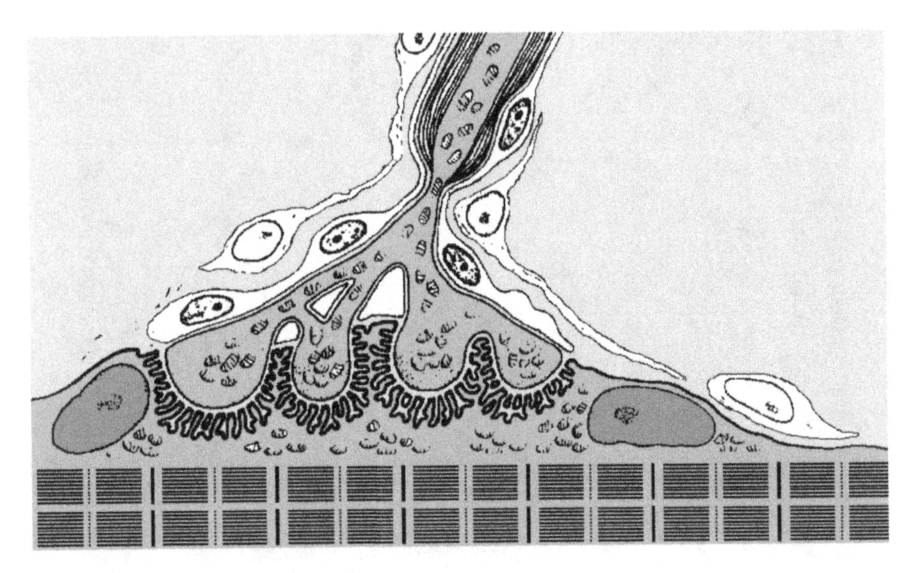

FIGURE 8: Neuromuscular junction. Permission pending.

The Sarcomere

Each myofibril is organized segmentally in contractile units termed sarcomeres. These are apparent in a low-power transmission electron micrograph of a human muscle biopsy (Figure 9). Note the myonucleus (N), the basal lamina (BL), the capillary (BV) in the endomysium, and the sarcomere bands. The A- (anisotropic), I- (isotropic), and Z-bands are evident. At higher magnification (Figure 10), more of the sarcomeric structure can be seen. Each sarcomere is delimited by the Z-bands (also termed the Z-disc). The A-bands are in the center of the sarcomere and are exactly 1.6 μm long in vertebrates. The major protein within the A-band is myosin, and it is the length of the myosin filaments (also termed the thick filaments) that establishes the length of the A-band. The I-bands extend from the Z-bands to the ends of the A-bands. The major protein of the I-band is actin, organized into F-actin filaments (thin filaments). Thin filaments are anchored at one end to the Z-discs. [The Z-band end of the thin filament is termed the barbed end; the opposite end of the thin filament is termed the pointed end. This terminology comes from the decoration of thin filaments with myosin S1. It need not concern us here, but its importance comes from the fact that it proves that actin filaments are polarized in opposite directions on either side of each Z-band. As you may recall, the barbed end is the plus end of the thin filament. The barbed end is also termed the plus end and the pointed end is the minus end, indicating the principal sites of G-actin addition or removal from the filament polymer.] Upon shortening, the I-bands decrease in length and the Z-bands move toward the A-bands. The length of the A-bands does not change until the Z-bands actually crunch the tips of the thick filaments, causing contraction bands. Muscle contraction occurs by the sliding of thin filaments between thick filaments. The total contraction of a muscle fiber is the summated contractions of each sarcomere, which in turn reflects the longitudinal interdigitation of thin and thick filaments (Figure 11). This mechanism is termed the sliding model of muscle contraction and is now universally accepted as the structural basis of striated muscle contraction.

A more complete understanding of the structural basis of muscle contraction requires a discussion of the other proteins in the sarcomere. The most important is the protein titin, which forms a third filament in the sarcomere termed the elastic or titin filament. Titin is the largest protein thus far uncovered in the animal and plant kingdoms. Its mass is ~300 MDa and single molecules actually span the distance from the Z-band to the M-band in the middle of the A-band (see Figure 12). Thus, single titin molecules are more than 1 μm long. Titin is tightly anchored to the thick filament, but in the I-band, it forms an elastic spring. It accounts for the series elastic component of muscle and also for the fact that a sarcomere overextended beyond thick and thin filament overlap will return to its native structure when longitudinal stretch is released. When titin is destroyed, this elasticity disappears. Another set of proteins exist in the thin filament. These include tropomyosin and troponin (discussed below), nebulin (another large protein of ~700 kDa), tropomodulin, and CapZ. The combination of nebulin and tropomodulin is important in the length regulation of the thin filaments. CapZ appears to be important in the anchorage of thin filaments into the Z-band and linking thin filaments with the major protein of the Z-band, α-actinin. Tropomodulin is required for terminating the longitudinal growth of thin filaments.

Note in Figure 12 that the thick filaments exhibit a series of lateral projections extending toward the thin filaments. These projections or cross-bridges are the heads of myosin molecules, and contain the actin-binding and ATPase domains of myosin.

The tails of the myosin molecules contain the polymerization domain of the molecule necessary for aggregation of myosin molecules to form the thick filaments. Note also that each half of the thick filament is oppositely polarized. The M-band represents the center of the thick filament and myosin is oriented in opposite directions on each half of the A-band.

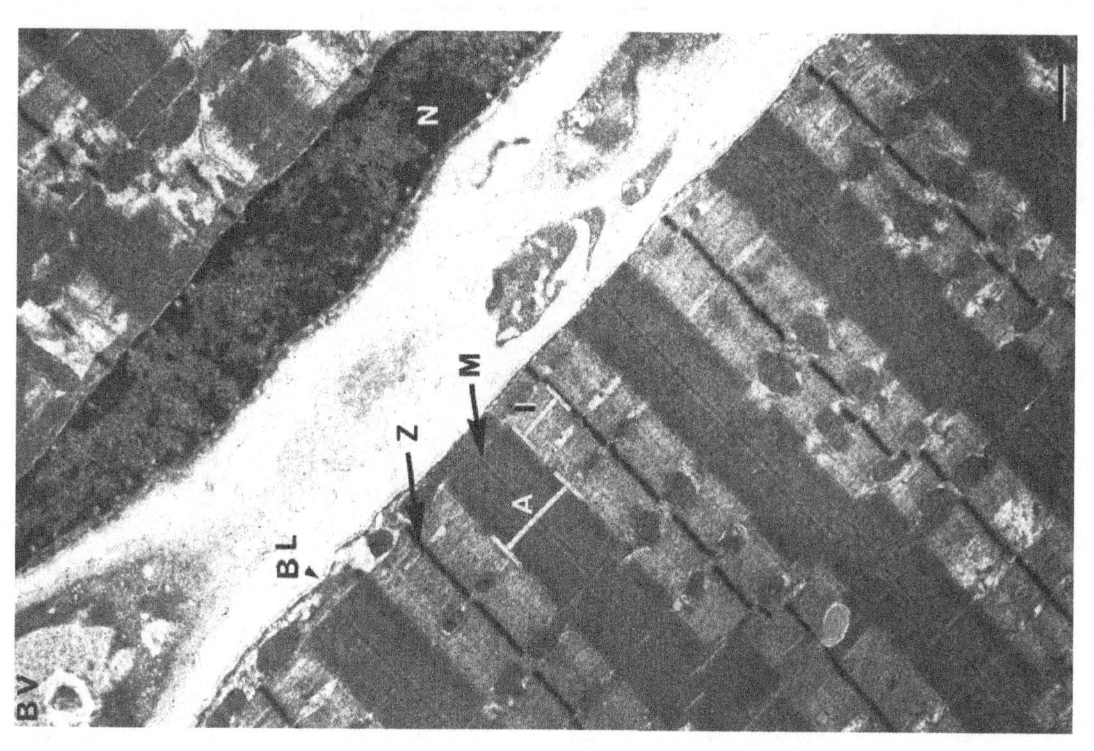

FIGURE 9: Electron micrograph of biopsied human vastus lateralis muscle (normal). Permission pending.

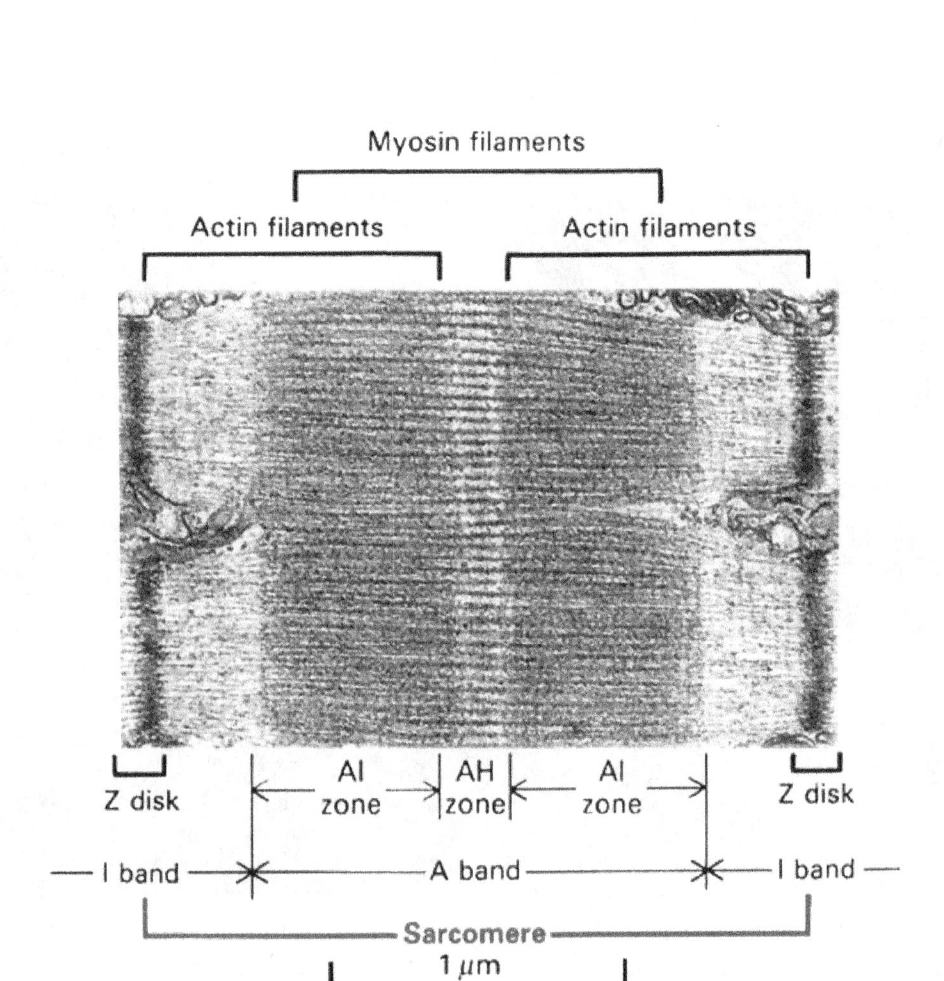

FIGURE 10: Saromeric structure seen at higher magnification. Permission pending.

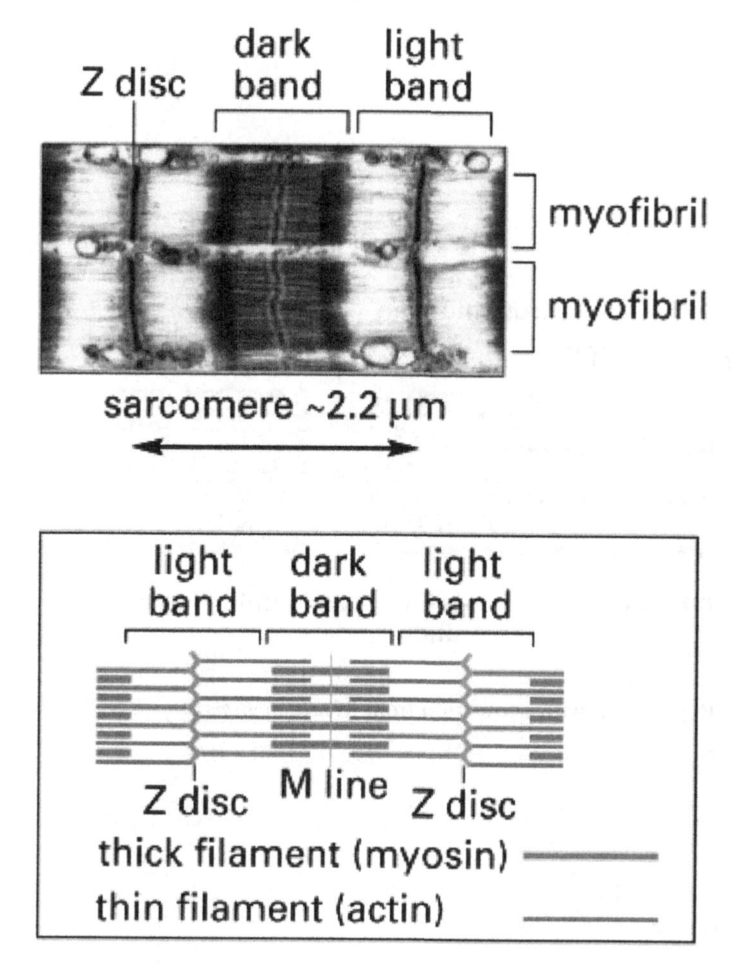

FIGURE 11: Reproduced with permission from *Molecular Biology of the Cell*, 4th edition, Garland Science Publishers.

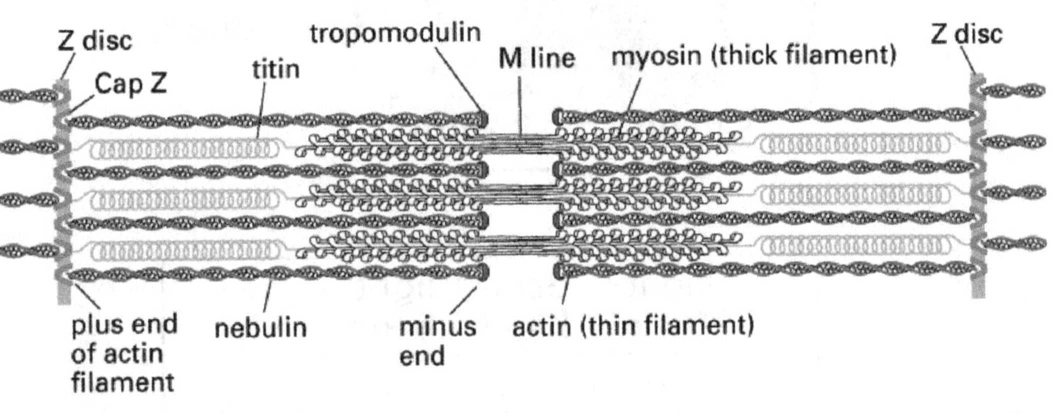

FIGURE 12: Reproduced with permission from *Molecular Biology of the Cell*, 4th edition, Garland Science Publishers.

The Myosin Molecule

Figure 13 shows an electron micrograph of rotary shadowed single myosin molecules. Note that each molecule has two globular heads and a long flexible tail. The globular heads contain the actin-binding sites and the ATPase catalytic site. Each head is termed the S1 fragment of myosin. Myosin S1 has been crystallized and its high resolution structure been established by X-ray crystallography (Figure 14).

The S1 portion of myosin has two major sections: the motor domain and the neck or lever arm. We now know that movement of actin filaments is caused by a rotary swing of the neck, while the motor domain is tightly bound to actin. The tail of myosin is not shown on this figure; it would run vertically on the right of Figure 14. A computerized reconstruction of myosin S1 bound to the actin filament is shown in Figure 15. A complete crystal structure of this complex has been not completed, but the model is based on the independent crystal structures of F-actin and myosin S1.

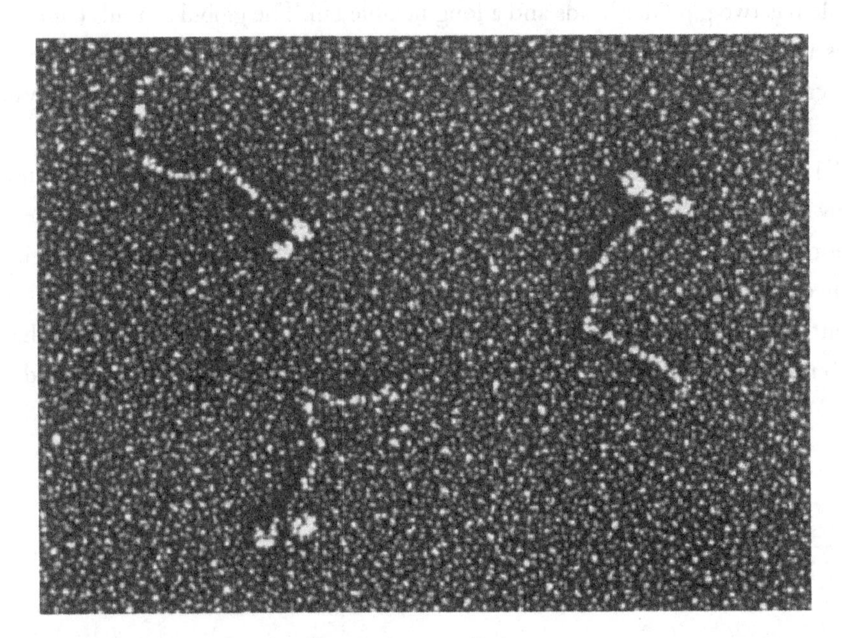

FIGURE 13: Single myosin molecules. Permission pending.

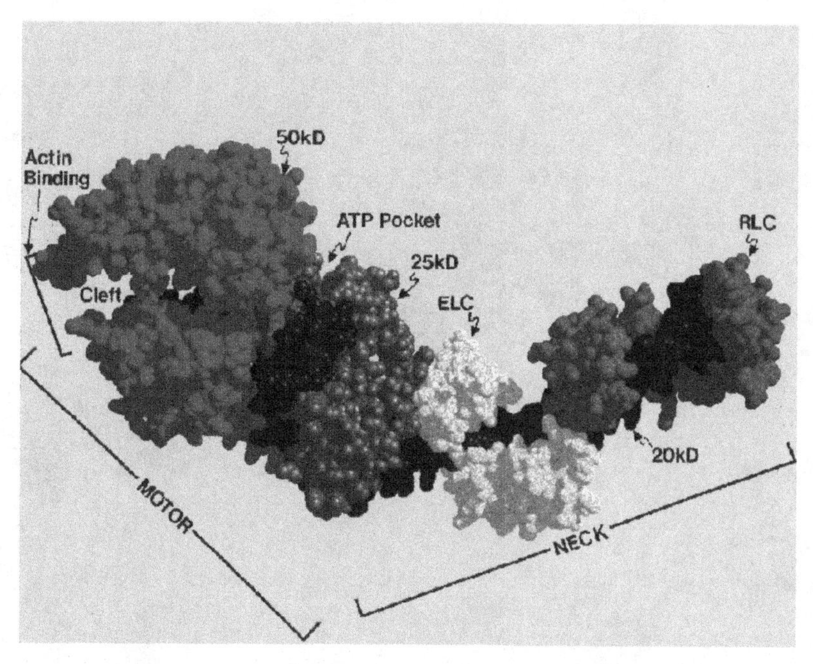

FIGURE 14: Myosin S1. Permission pending.

FIGURE 15: Crystal structure of myosin head linked to actin filament. Permission pending.

Regulation of Muscle Contraction

The thin filament accessory proteins tropomyosin and troponin (Figure 16) are essential for the Ca^{2+} regulation of striated muscle contraction. In the resting muscle cell, sarcoplasmic Ca^{2+} levels are very low ($<10^{-7}$ M). Upon excitation of muscle (see Figure 17), Ca^{2+} levels rise in the sarcoplasm by Ca^{2+} release from the SR and influx from the extracellular space. This elevates sarcoplasmic levels to $>10^{-6}$M. This results in the binding of Ca^{2+} to troponin, which in turn causes a conformational change of tropomyosin. This change in tropomyosin causes an activation or release of a steric block to the myosin binding site on actin (Figure 17). The end result is the rapid binding of myosin to actin and contraction of the sarcomere. In short, Ca^{2+} releases a block to contraction by tropomyosin. The precise mechanisms of this reaction are still under intense study.

Our current understanding of how ATP hydrolysis drives muscle contraction is shown in Figure 18, the cross-bridge cycle. In the absence of ATP myosin binds strongly to actin. This would represent the end of a preceding power stroke of the cross-bridge. Addition of ATP (in the presence of Ca^{2+}) detaches the cross-bridge (i.e., dissociates actin from myosin) and the cross-bridge then swings to its next "cocked" position. ATP hydrolysis occurs, but ADP and Pi remain bound to myosin. In the next step, Pi is released and the cross-bridge binds to actin and forced generation begins. Upon ADP release, the power stroke occurs and we return to the first position. This figure is not exactly correct. The cross-bridge does not move; it is the lever arm at the neck of myosin that undergoes major conformational changes.

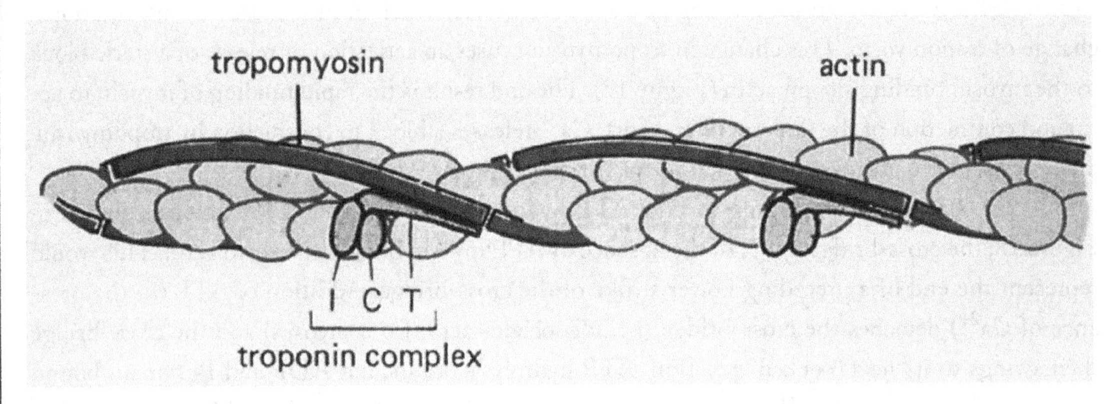

FIGURE 16: Model of the tropomyosin-troponin complex with actin. Permission pending.

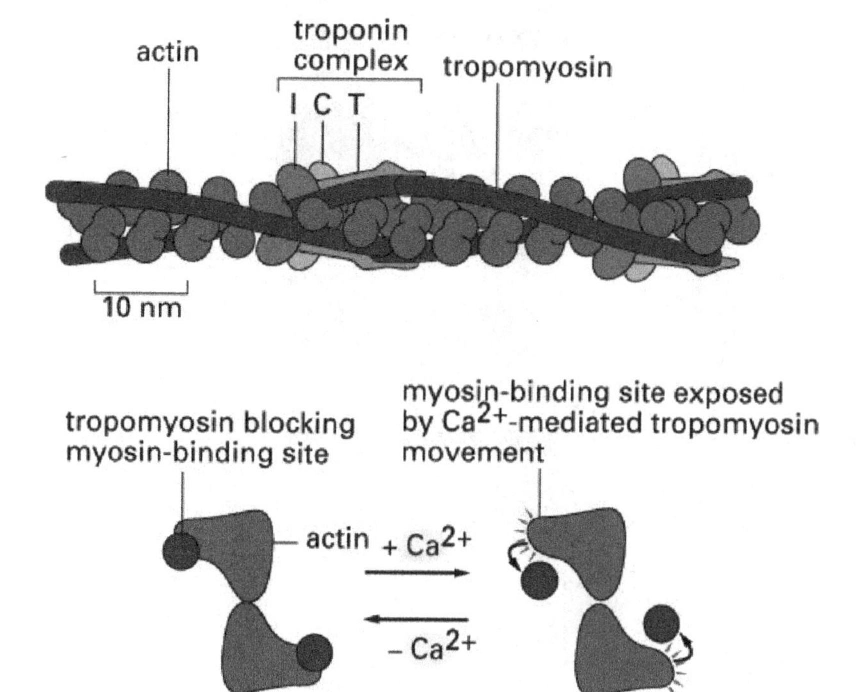

FIGURE 17: Reproduced with permission from *Molecular Biology of the Cell*, 4th edition, Garland Science Publishers.

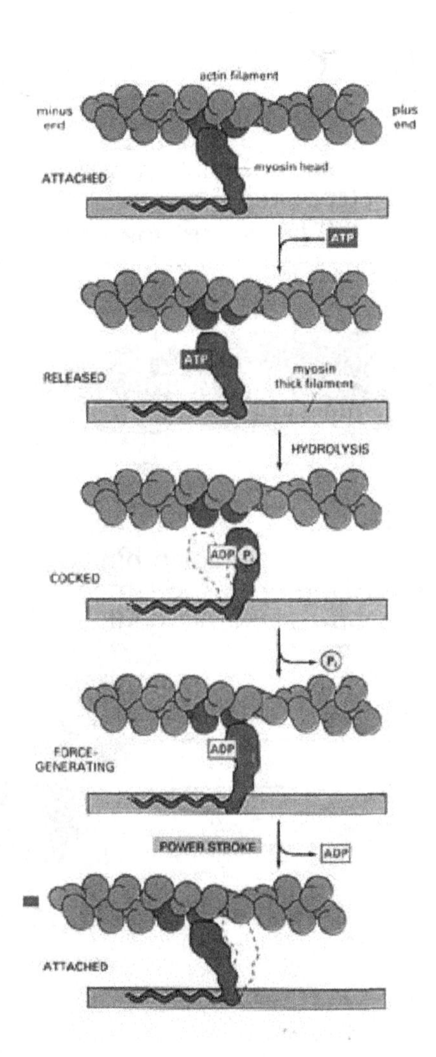

FIGURE 18: Binding of myosin to actin in relation to ATP hydrolysis. Permission pending.

The Sarcoplasmic Reticulum

Skeletal and cardiac muscles possess a unique smooth endoplasmic reticulum termed the sarcoplasmic reticulum. This membrane system is essential for Ca^{2+} regulation in striated muscle. The SR envelops the myofibrils and is in direct contact with invaginations of the sarcolemma termed the transverse tubules (T-tubules) (Figure 19). A specialized membrane junction is found at the sites where the SR contacts the T-tubules; it is termed the triad, since two elements of the SR contact a single T-tubule. It is in the T-tubules that action potentials from the sarcolemma reach the interior of the cell and, at the triad, trigger release of Ca^{2+} from the SR. The elaboration of the SR is in direct proportion to the speed of contraction and relaxation of a muscle. For example, the bat cricothyroid muscle, which is responsible for sonar echo location of bats, contains a remarkably well developed SR.

An electron micrograph of the triad is shown in Figure 20. Labels indicate the site of the dihydropyridine channel where Ca^{2+} release into the sarcoplasm is initiated. A diagram of the events at the triad is presented in Figure 21.

A voltage-sensitive protein channel at the triad (dihydropyridine channel) initiates Ca^{2+} release into the SR. This in turn leads to the massive release of Ca^{2+} into the sarcoplasm (cytosol).

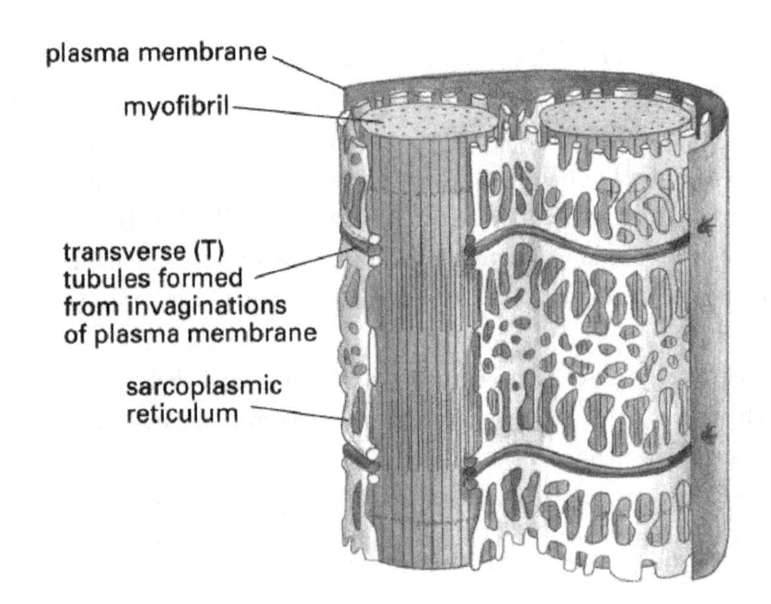

FIGURE 19: Reproduced with permission from *Molecular Biology of the Cell*, 4th edition, Garland Science Publishers.

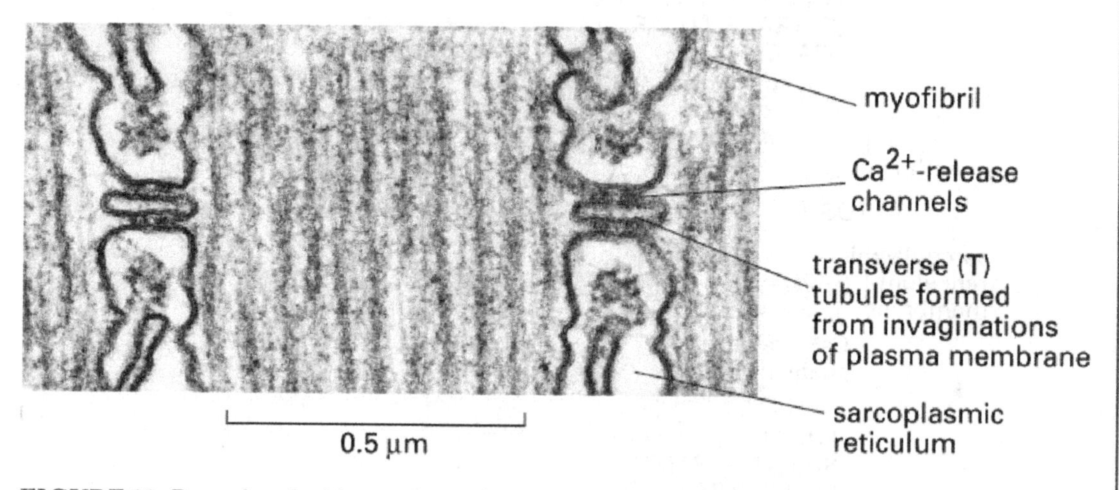

FIGURE 20: Reproduced with permission from *Molecular Biology of the Cell*, 4th edition, Garland Science Publishers.

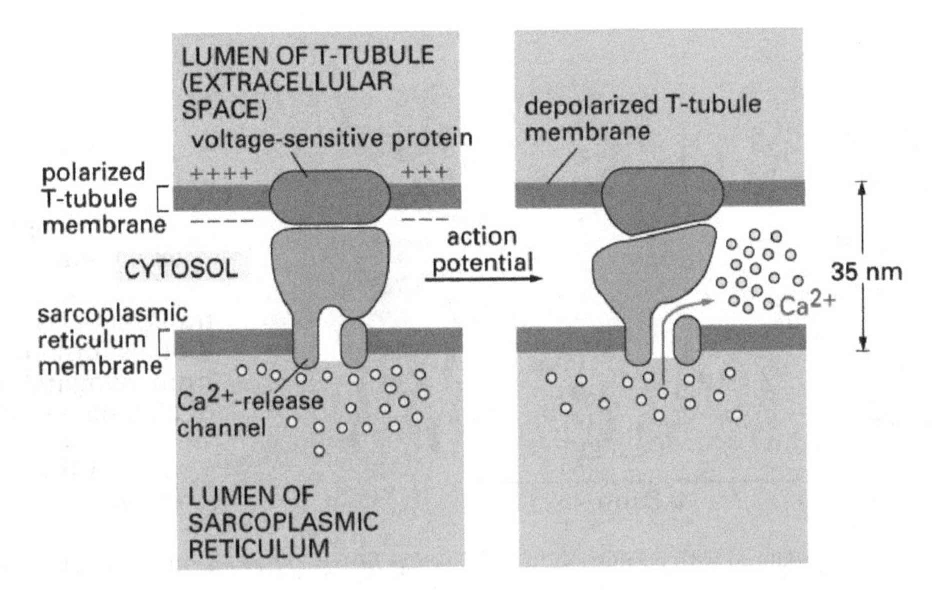

FIGURE 21: Reproduced with permission from *Molecular Biology of the Cell*, 4th edition, Garland Science Publishers.

Muscle Fiber Types

All human skeletal muscles exhibit variability in the morphology, biochemistry, and physiology of their myofibers. For example, some muscles are designed to contract very rapidly (e.g., extraocular muscles), while others are more designed to sustain posture (e.g., strap muscles of the back). Cardiac muscles must contract ~70 times each minute and must do so for an entire lifetime. This variation in function is reflected in the contractile isoforms expressed in each muscle, the number of mitochondria found in the cells, the extent of glycogen accumulation, and the cross-sectional diameter of the fibers. There are many ways to classify these differences. A simple classification that works fairly well is to divide the muscles into fast-twitch glycolytic fibers, fast-twitch aerobic fibers, slow-twitch aerobic, and slow tonic fibers. The names indicate the relative speeds of contraction and their relative dependence on oxidative phosphorylation or glycolysis for ATP generation. Figure 22 illustrates fiber type differences in mitochondrial content demonstrated by staining for succinic dehydrogenase.

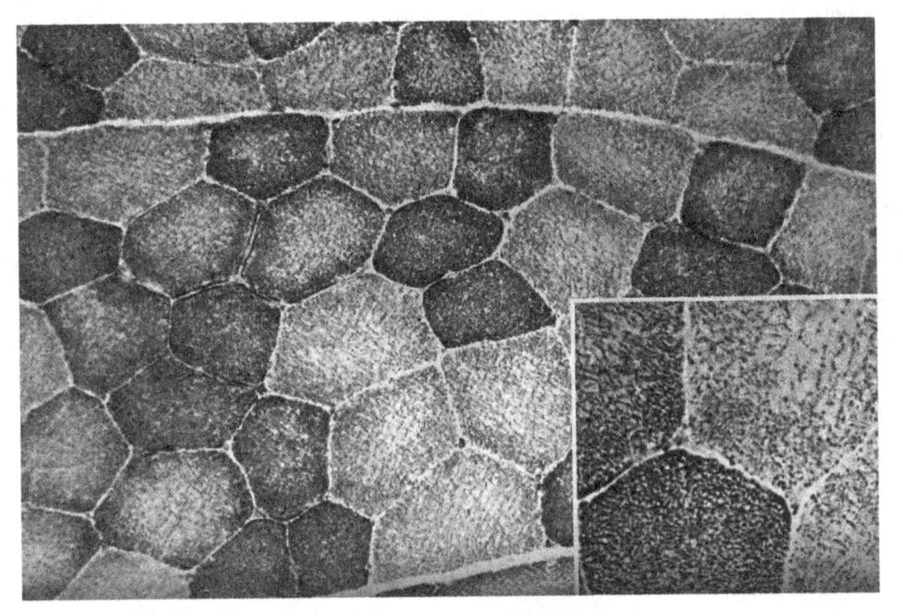

FIGURE 22: Succinic dehydrogenase staining of muscle to exhibit fiber type differences in mitochondrial content. Permission pending.

The Myotendon Junction

At the ends of the myofiber, there is a specialized structure to attach the muscle cell to collagenous fibers. This is termed the myotendon junction. It is illustrated in Figure 23. The arrowheads in the figure point to an elaborate series of deep invaginations of the sarcolemma into which collagen fibers insert. The sarcolemma exhibits a modified adherens-like junction at this site, with an extensive network of parallel actin filaments attached between the myofibrils and the adherens-like junction. Figure 24 shows this in an electron micrograph.

The deep invaginations of the muscle cell surface increase its surface area many fold, thus providing a larger surface for adhesion.

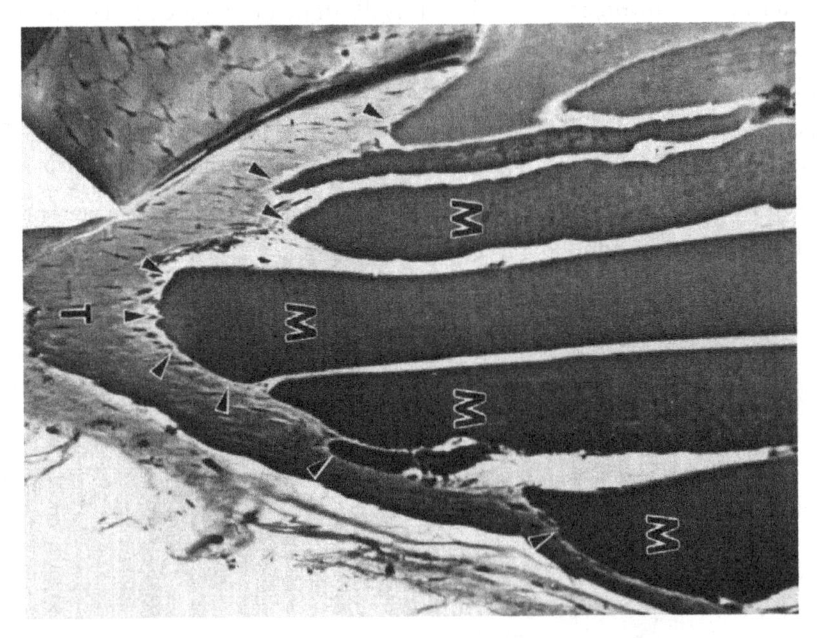

FIGURE 23: Light micrograph of the myotendon junction. Permission pending.

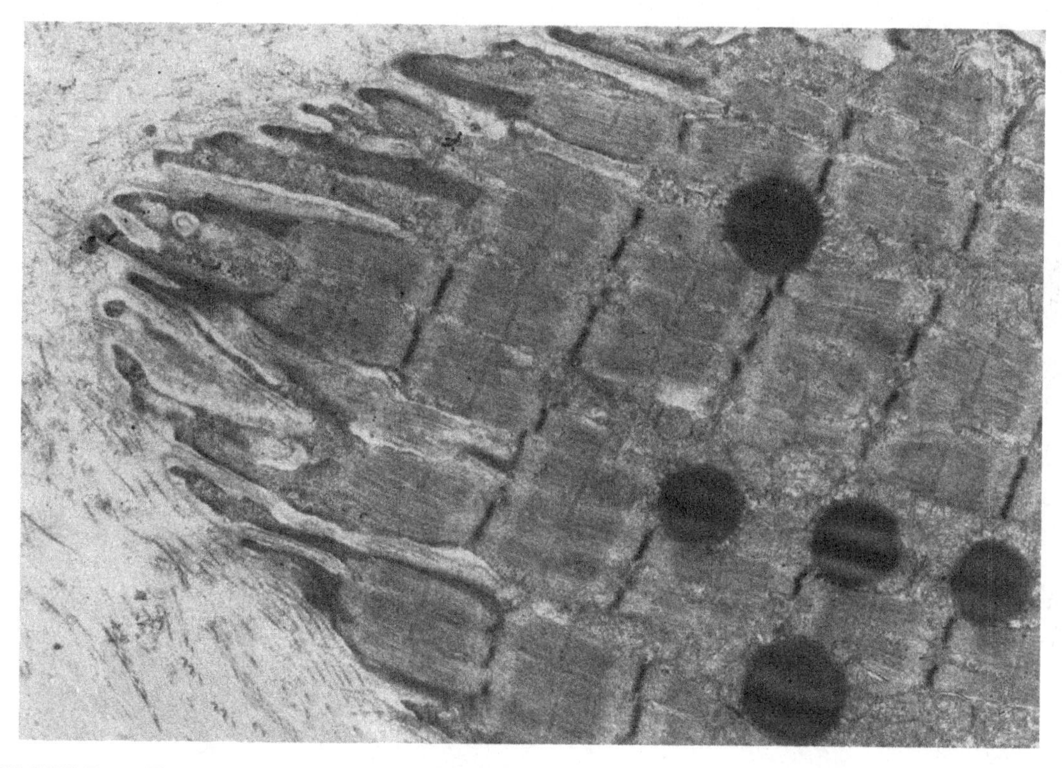

FIGURE 24: Transmission electron micrograph of a myotendon junction. Permission pending.

The Satellite Cell

Skeletal muscle contains a unique stem cell population termed the satellite cell. This cell was discovered by Alex Mauro in the 1950s working at Rockefeller University while studying the physiology of frog muscle. It has since turned out that this population of cells is responsible for the regenerative potential of skeletal muscle after injury. Satellite cells are found beneath the basal lamina in direct contact with the muscle fiber plasma membrane (Figure 25). In the adult, these cells exist in an arrested G_0 state, but upon injury to muscle, the cells reenter the mitotic cycle and create a substantial pool of myoblasts that are capable of both fusing with the injured fibers or forming new fibers de novo by the embryonic process termed myogenesis.

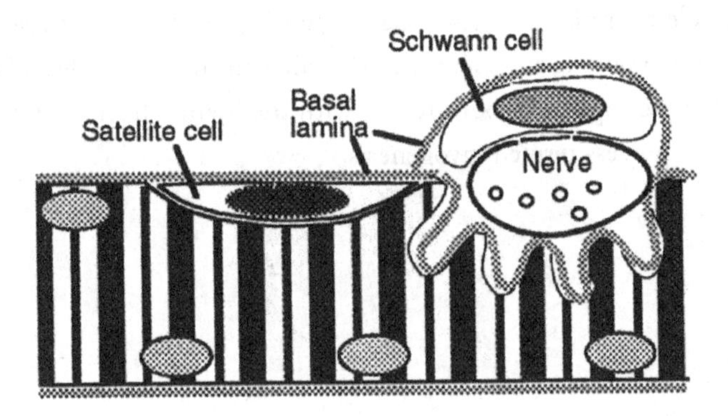

FIGURE 25: Diagram showing the position of satellite cells and Schwann cells in relation to the basal lamina at the myoneural junction. Permission pending.

Muscular Dystrophy

Over the past 20 years, there has been an intensive study of the pathophysiology and genetics of the often-fatal human muscular dystrophies. These are usually heritable diseases that predominantly affect skeletal muscle, although some of them also affect the heart. Death usually results from respiratory failure secondary to the degeneration of diaphragmatic muscles and/or heart failure. The most tragic is Duchenne muscular dystrophy, an X-linked dystrophy in which mothers carry the mutant gene but the disease manifests in boys who inherit the mutant X chromosome. In a remarkable series of studies conducted at Harvard by Louis Kunkel and colleagues, the mutant gene was identified and sequenced. It was found to encode a large protein now termed dystrophin (Figure 26). It is part of a submembranous complex of proteins termed the dystrophin complex. This protein (or others in the complex), when mutated, causes a fragility of the sarcolemma, leading to muscle degeneration probably secondary to the leakage of Ca^{2+} into the muscle sarcoplasm. Characteristically, the dystrophic muscle exhibits degeneration, regeneration, and lipid deposition in adjacent areas of the tissue. Early in the disease, degeneration and regeneration predominate, but eventually, the degenerated muscle cannot be repaired and adipose tissue replaces the muscle. It is thought that eventually satellite cells exhibit mitotic senescence and lose their ability to undergo myogenesis. In Figure 27, a boy with Duchenne muscular dystrophy exhibits characteristic difficulties in attempting to stand erect. This is known as Gower's sign (or Gower's stance) and is diagnostic of early stages of the disease. As you may know, most cases of this disease result in premature death, typically in late teens or early adulthood. Definitive diagnosis requires genetic testing and usually involves immunofluorescent staining of biopsied leg muscle (vastus lateralis) to assess the presence or absence of dystrophin. Other forms of muscular dystrophy, most of which are not X-linked, may involve other proteins of the dystrophin complex.

Finally, the connection between muscle wasting and dystrophin can be clearly discerned from histological examination of biopsied leg muscle (Figure 28). The absence of functional dystrophin at the plasma membrane of the muscle cells weakens the membrane to the extent where normal exercise tears the membrane, allowing influx of extracellular ions such as Na^{+} and Ca^{2+}, which destroy sarcoplasmic function. Affected muscle tissue loses contractile cells that are replaced with extracellular connective tissue and fat cells (adipocytes).

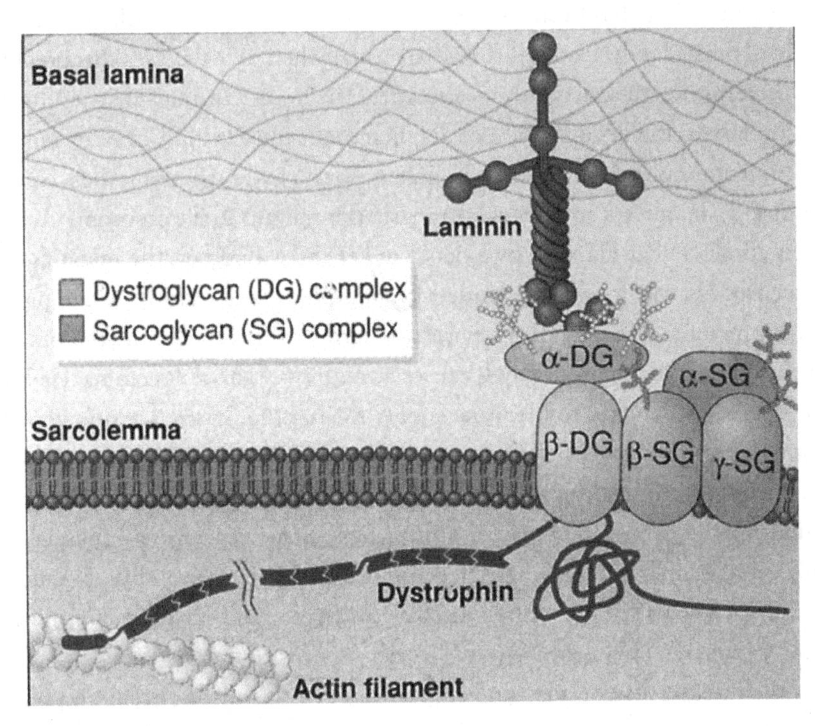

FIGURE 26: Dystrophin. Permission pending.

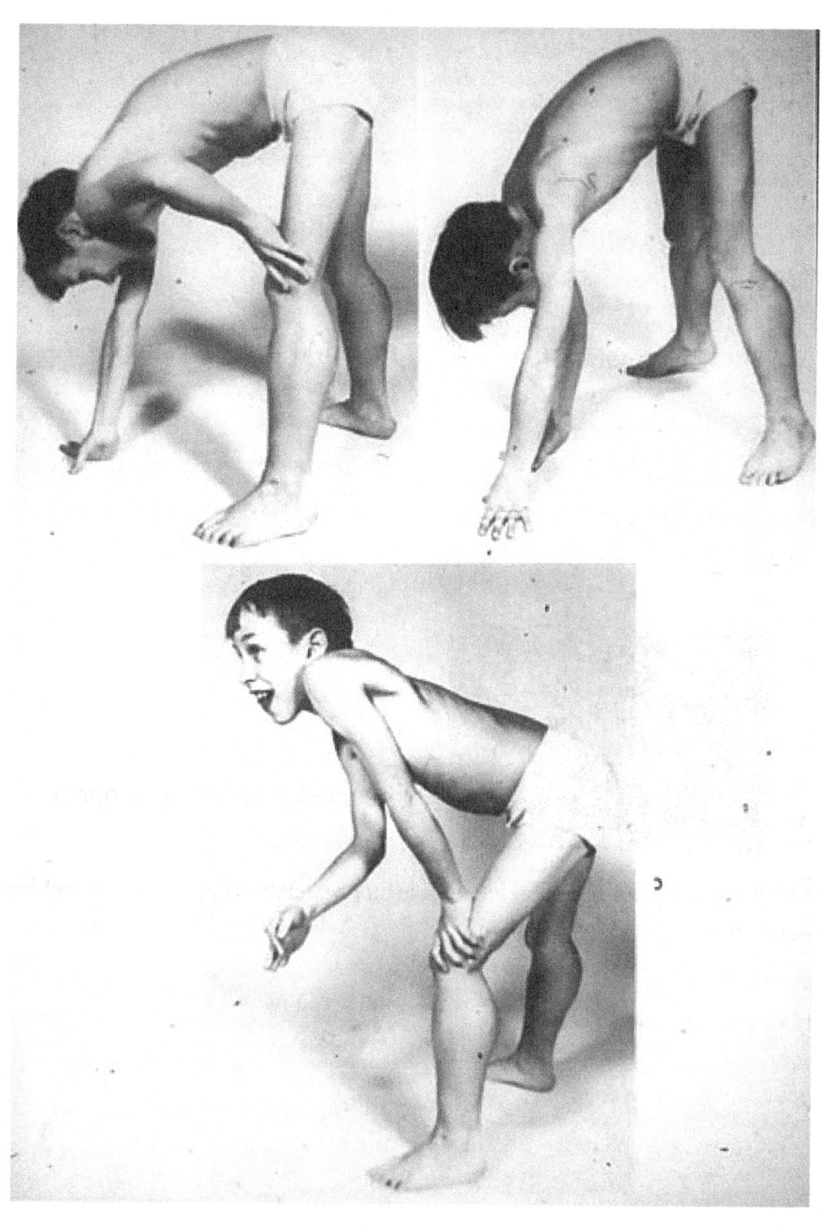

FIGURE 27: Gower's sign in a boy with Duchenne muscular dystrophy. Permission pending.

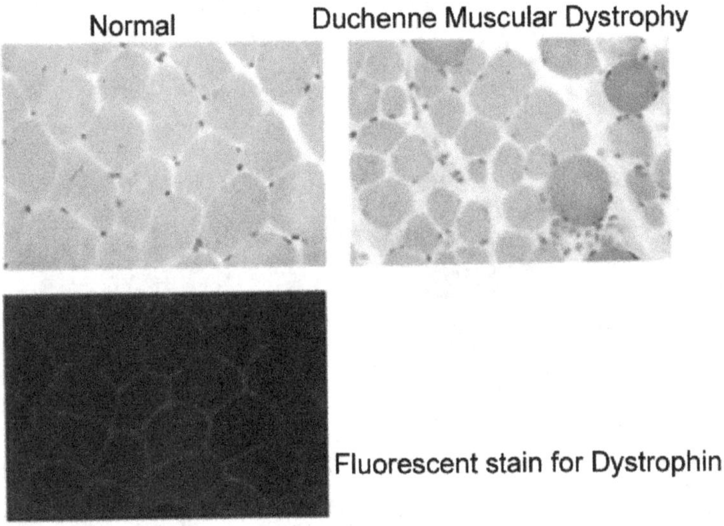

Normal

Duchenne Muscular Dystrophy

Fluorescent stain for Dystrophin

FIGURE 28: Histological examination of leg muscle biopsy: relationship of muscle wasting and dystrophin. Permission pending.

Recommended Readings

Alberts et al., *Molecular Biology of the Cell*, 4th ed., pp. 961–965.
Ross et al., *Histology*, 4th ed., pp. 246–281.

Series Editor Biography

Joel D. Pardee, PhD is an Associate Professor of Cell and Developmental Biology at Weill Cornell Medical College. He served as the associate dean of the Weill Cornell Graduate School of Medical Sciences of Cornell University from 1997 to 2007, as the associate dean for Research Services from 2007 to 2009, as the director of the Office of Postdoctoral Affairs from 2005 to 2009, and is the current director of student research. He has been active in medical education for 25 years, both as a lecturer and as a course director of the first year medical school course, molecules, genes, and cells, and has been recognized with several excellence in Teaching Awards by the Weill Cornell's medical student body. Dr. Pardee is also the founder of Neural Essence, LLC, and the author of more than 50 research papers and patents in the fields of cell motility, cancer cell metastasis, and metabolic enhancement therapy.

Index

Colloquium Series on the Cell Biology of Medicine

Editor
Joel D. Pardee, Weill Cornell Medical College

Forthcoming Books (continued from page iii)

Cancer Genetics
A.M.C. Brown
2009

Cancer Invasion and Metastasis
A.M.C. Brown
2009

Cartilage: Keeping Joints Functioning
Michelle Fuortes
2009

The Cell Cycle and Cell Division
Joel D. Pardee and A.M.C. Brown
2009

Cell Metabolic Enhancement Therapy for Mental Disorders
Joel D. Pardee
2009

Cell Motility in Cancer and Infection
Joel D. Pardee
2009

Cell Structure and Function: How Cells Make a Living
Joel D. Pardee
2009

Cell Transformation and Proliferation in Cancer
A.M.C. Brown
2009

Cholesterol and Complex Lipids in Medicine: Membrane Building Blocks
Suresh Tate
2009

Creating Proteins from Genes
Phil Leopold
2009

Development of the Cardiovascular System
D.A. Fischman
2009

Digestion and the Gut: Exclusion and Absorption
Joel D. Pardee
2009

The Extracellular Matrix: Biological Glue
Joel D. Pardee and A.M.C. Brown
2009

Fertilization, Cleavage, and Implantation
D.A. Fischman
2009

Gastrulation, Somite Formation, and the Vertebrate Body Plan
D.A. Fischman
2009

Generating Energy by Oxidative Pathways: Mitochondrial Power
Suresh Tate
2009

Heart Structure and Function: Why Hearts Fail
D.A. Fischman
2009

Hematopoiesis and Leukemia
Michelle Fuortes
2009

How Amino Acids Create Hemoglobin, Neurotransmitters, DNA, and RNA
Suresh Tate
2009

The Human Genome and Personalized Medicine
Phil Leopold
2009

Mechanisms of Cell & Tissue Aging: Why We Get Old
Joel D. Pardee
2009

Mendelian Genetics in Medicine
Phil Leopold
2009

Metabolism of Carbohydrates: Glucose Homeostasis in Fasting and Diabetes
Suresh Tate
2009

Metabolism of Fats: Energy in Lipid Form
Suresh Tate
2009

Metabolism of Protein: Fates of Amino Acids
Suresh Tate
2009

Neurulation, Formation of the Central and Peripheral Nervous Systems
D.A. Fischman
2009

Non-Mendelian Genetics in Medicine
Phil Leopold
2009

Skeletal Muscle, Muscular Dystrophies, and Myastheneis Gravis
D.A. Fischman
2009

Skin: Keeping the Outside World Out
Joel D. Pardee
2009

Stem Cell Biology
A.M.C. Brown
2009

Therapies for Genetic Diseases
Phil Leopold
2009

Tissue Regeneration: Renewing the Body
Joel D. Pardee
2009

Colloquium Series on Integrated Systems Physiology

Editors

D. Neil Granger, Louisiana State University Health Sciences Center

Joey Granger, University of Mississippi School of Medicine

Forthcoming Books

Capillary Fluid Exchange
Ronald Korthuis
2009

Endothelin and Cardiovascular Regulation
David Webb
2009

Hemorheology and Hemodynamics
Giles Cokelet
2009

Homeostasis and the Vascular Wall
Rolando Rumbaut
2009

Inflammation and Circulation
D. Neil Granger
2009

Integrated Cardiovascular Responses to Exercise
Doug Seals
2009

Liver Circulation
Wayne Lautt
2009

Lymphatics
David Zawieja
2009

Ocular Circulation
Jeffrey Kiel
2009

Pulmonary Circulation
Mary Townsley
2009

Regulation of Arterial Pressure
Joey Granger
2009

Regulation of Cardiac Contractility
John Solaro
2009

Regulation of Endothelial Barrier Function
Harris Granger
2009

Regulation of Tissue Oxygenation
Roland Pittman
2009

Regulation of Vascular Smooth Muscle Function
Raouf Khalil
2009

Skeletal Muscle Circulation
Ronald Korthuis
2009

Vascular Biology of the Placenta
Yuping Wang
2009

Colloquium Series on Developmental Biology

Editor
Daniel S. Kessler, University of Pennsylvania

Forthcoming Books

Fibroblast Growth Factor Signaling
Elizabeth Pownall and Harvey Isaacs
2009

Formation and Differentiation of Placodes
Jean-Pierre Saint Jeannet
2009

Formation of the Embryonic Mesoderm
Daniel Kessler
2009

Maternal Control of Embryogenesis
Florence Marlow
2009

Neural Crest Lineage
Patricia Labosky
2009

Organization of the Nervous System
Dale Frank
2009